Merry Christmas "Grandma" '96

Love Terry

D0688536

BARNS

BARNS

Nicholas S. Howe

MetroBooks

MetroBooks

An Imprint of Friedman/Fairfax Publishers

© 1996 by Michael Friedman Publishing Group, Inc.

All rights reserved. No part of this publication may be reproduced,
stored in a retrieval system, or transmitted, in any form or by
any means, electronic, mechanical, photocopying, recording, or
otherwise, without prior written permission from the publisher.

Library of Congress Cataloging-in-Publication data available upon request.

ISBN 1-56799-290-0

Editor: Stephen Slaybaugh
Art Director: Lynne Yeamans
Designer: Susan E. Livingston
Photography Director: Christopher C. Bain

Color separations by Fine Arts Repro House Co., Ltd.
Printed in Hong Kong and bound in China by Midas Printing Limited

For bulk purchases and special sales, please contact:
Friedman/Fairfax Publishers
Attention: Sales Department
15 West 26th Street
New York, NY 10010
212/685-6610   FAX 212/685-1307

ACKNOWLEDGMENTS

Many excellent books have been made about barns, and I am
indebted to the authors whose books I used to confirm my
childhood memories and widen my current views, especially
Eric Sloane, who has created a delightful series about an
American way of life that is slowly being lost. Mr. Sloane is the
one who recovered the advice given to a colonial carpenter that
appears as the preface to this book.

CONTENTS

~

Rockville, IN

In the art of building, an intelligent and expert carpenter is entitled to the foremost place, or first degree of eminence; for he is able to erect a house without calling in either bricklayer or mason, neither of whom can execute the like task independent of his assistance. His profession depends on the practical application of the most plain, simple, and unerring principles; and more pleasure results from the view, as well as more comfort from the use, of a neat well-constructed common house, than from the most superb but ill-contrived palace, where fanciful ornaments are frequently introduced with no better intention than to disguise blemishes in proportion and symmetry. Strength and convenience are the two most essential requirements in building; the due proportion and correspondence of parts constituting a beauty that always first attracts the eye; and where that beauty is wanting, carving and gilding only excite disgust.

—*The Carpenter's Pocket Dictionary*, 1797

GROWING UP WITH BARNS

My grandfather was born in 1875, and when I was a child he showed me how to work a flail. The flail is a farming tool that has not changed in all of human memory. It's simple enough to look at—just two sticks held together by a flexible joint—but it's very difficult to use properly. It is what was used to separate the wheat from the chaff in biblical times, and although my grandfather was not a farmer, he'd learned how to manage a flail when he was a boy.

My grandfather also told me that you could tell a wheat farmer's barn just by looking at it; there would be large doors on each side of the work floor so that the breeze would blow the chaff away when it was tossed into the air. I did not learn how to manage a flail. Our family had not been farmers since they lived on the western frontier, which meant that they lived on the Connecticut River north of the Massachusetts colony. I knew that in those days people and animals would often stay in the same building so the animals' heat would help keep the people warm in winter. A goat was also put in with the cows. Goats give off an unusual amount of heat, and this had a calming effect on the larger animals and helped them rest better.

All that experience had receded into ancestor stories by the time I was a child, but it must have left its mark because my favorite places to play were barns. We lived in Deerfield, Massachusetts, one ridge west of the Connecticut River. The town was one mile (1.6km) long and it had three schools and three farms in it, with another school and another farm on the mountain above us. The members of my family were teachers, but almost everyone I knew in grammar school lived on farms.

The closest farm to our house was four doors to the south, where I believed the Wells family had been at work since there was still danger of Indian raids. This longevity invested the Wells farm with

OPPOSITE· *Washington*

~

11

a distance that we could never quite bridge, and we did not go into their barns. It was enough to look over the fence into the wallow where the pigs would sprawl, grunting and nosing in the mud for any corn cobs they might have forgotten. We were on closer terms with the horses. In the winter, they pulled a snowplow along the sidewalk that ran through our front lawns while steam rolled off their flanks and trailed away behind them. The plow was a wooden wedge with faded red paint, with a board across the vee for one of the Wells men to stand on as he held the reins. This was heroic work, which made sense to me; the main barn on the Wells farm was the largest structure most of us had ever seen. It established the far end of the scale that we used to measure what men could do.

The Cowles' farm was at the north end of town and it was more accessible because their daughter and son were the same age as my sister and I. Their barn was mysterious, but it was a mystery that yielded to children's curiosity. There were dark, narrow, unexplained passages running around the back of the bottom level of animal pens that were barely wide enough for us to slide through; it was almost like being in the pen with the animals, but still safe. The top of the barn was wide and airy, almost like being outdoors. A skeleton of beams laced through these lofty spaces, and we could climb up there and dive into the hay.

The bull was kept on the main floor of the barn and his pen radiated danger. The pen was reinforced with timber and heavy planks leaving barely enough space for us to look through at the brute inside; on our best days, we'd use the spaces like a ladder and look in over the top bracing. This was the only animal that we were really afraid of, and we were right to be so: the power of a bull could not be predicted or restrained, and anything might happen.

If proximity to the bull defined the limits of bravery, we learned about temptation at the silo. This tower was filled with chipped corn, which was blown in at the top by a complex machine, and the bottom was always wet. We understood that if we pulled a handful of silage out of the openings at the bottom and chewed on it, we'd get drunk. No one I knew had ever dared do this, but we had the news on good authority.

OPPOSITE: *Central Missouri*

Sin also lurked in the meadows outside town. Deerfield was a sort of island in a river valley filled with tobacco fields, and after the crop was made the leaves would be hung in long, narrow, single-level barns. All the other barns were like castles, great strong places with each of their many parts devoted to a different purpose in the management of crops, feed, animals, and equipment. Tobacco barns, however, were empty most of the year and seemed insubstantial. Their siding was made of wide planks, each as long as the space between the ground and the roof, and every second plank was hinged at the top. After the tobacco was cut and hung, these siding planks would be propped out at the bottom so air could circulate and dry the leaves. So, during their one chance at usefulness, tobacco barns looked like a venetian blind that hadn't quite worked out right, and they rattled when the wind blew. But we knew what tobacco was for, and at the time when a child tries smoking for the first time, that's where we went. The results were terrible.

Barns also taught us all we knew of death. One of my friends lived a mile south of town and his father raised chickens, a messy rattle-brained population that completely filled a two-story henhouse. At intervals, my friend's father would kill several of the chickens. He'd grab one of them by its feet, turn it upside down into the top of a metal funnel, and pull its head out the bottom. Then, with a quick motion, he'd slide a thin knife into the chicken's throat, and blood would run out its mouth like water from an opened faucet.

The Spruyt family had the farm on the ridge above town, at the very outer edge of walking distance. Actually, it was farther away than that; it was more like another world. Frick Spruyt was Dutch, and his American wife Antoinette seemed even more Dutch than he was. A man named Trask lived at the far end of the fields and helped with the heavy work. I always thought his name was probably "task" misspelled.

The way into the main house led through a room with racks of strange tools and devices, which, we understood, were used in animal husbandry. The barn was a vast place of cool darkness lit only by thin shafts of light coming in through the chinks; there was a haymow several stories high, and each kind of animal had its own section of the barn. The goats, for instance, had a long, narrow wing with many small windows and immaculate pine-paneled stalls, each topped by a brass fixture that held a slip of paper printed with the name of the resident.

The Spruyts were doing research with Togganberg goats, trying to develop a strain that would produce maximum butterfat for minimum feed back home in Holland, which was at that time occupied by the Germans and having hard times. Those name slips were important, because Togganberg goats tend to look very much alike.

Then came a day when one of the goats was testing the limits of her tether in search of something more exotic than the carefully controlled ration assigned to her. Before long, she found the slip of paper in the brass holder, and faithful to her species' reputation, she ate it. This news spread quickly, and by the next time one of the Spruyts came to check on the goats, anonymity was everywhere.

This sweet-smelling wing of the barn was, of necessity, for ladies only. Their collective other half were named Isaac and Abraham, immense brutes whose evil smell was exceeded only by their irascible tempers. They lived as far away and under as tight security as space and carpentry could provide. No friends of the tax man, the Spruyts let Isaac and Abraham out of their pen one day when Mr. Stebbins came to assess their property.

There was also a team of horses, as well as many ewes attended by rams with proud chests and massive curling horns. Our favorites were the oxen that pulled a high-wheeled blue cart; they were named Cyndric and Pindar, and they had long sweeping horns tipped with polished brass knobs.

On Christmas Eve we'd go up to the Spruyts for a traditional Dutch dinner. We ate in their large music room and Fred Hyde would be the only other guest. He was a gentle aesthete, a musician who specialized in the Baroque era and played the harpsichord and clavichord.

The Spruyts told us that once each year the animals could talk. This was at midnight on Christmas Eve, and the gift would only come if the right carols were sung, a different one for each kind of animal. After dinner we'd all walk out into the snow and the night, and go to the barn and sing to the goats and sheep and horses and oxen. Fred would play his clavichord, an ancient tinkling keyboard that he carried by a broad ribbon around his neck, and we'd make our way through the dark barn by the light of lanterns and sing a different carol at the stall of each animal. Then we'd leave quickly, because if any humans were around to hear, the gift of talking would not come.

PART I

FORM

Americawas an idea before it was a place. As soon as the European naviga-
tors discovered the vast new land across the ocean, strivers and dreamers of

every kind projected onto it their visions of a new world, a new access to

the lost Garden of Eden. Thus the garden became dominant as both fact

and symbol in America, the source of both virtue and rights, and few set-

tlers were as specific as Hector de Crevecoeur as he wrote from his new

world farm in 1782.

"Here the reflecting traveller might contemplate the very beginnings

and outlines of human society. Here we have in some measure regained the

ancient dignity of our species; our laws are simple and just, we are a race of

cultivators, our civilization is unrestrained, and therefore everything is

prosperous and flourishing. For my part I had rather admire the ample barn

of one of our opulent farmers, who himself falled the first tree of his plan-

tation, and was the first founder of his settlement, than study the dimen-

sions of the temple of Ceres."

LEFT: *Bozeman, MT*

~

17

At the time when the first European settlers were arriving in the New World, the most common roofing material available in the countryside was straw. Straw is what is left after the grain has been threshed out of wheat. It is similar to coarse hay and is no more waterproof than a handful of dry grass. But if straw is laid on a steeply pitched framework, water will run down the length of its stalk and drip off the end. If the straw is packed in bundles and laid on several layers deep, rain will run all the way down the pitch of the roof and drip off the eaves. This is the thatched roof, and it worked remarkably well—better than anything else that was available to the average man.

Europeans had built their roofs this way in the first New World settlements, but the abundant softwood forests enabled the settlers to start making shingles of wood. The

steep pitch of the thatched roof was retained, as it had been the key to a dry house; if a straw roof was flat, the rain would soak through rather than run off the stalk. This idea endured in the shingled roof. This simple gable roof formed the basis for barns from the time when the frontier was at the shore of the Atlantic, and whenever a sturdy building was needed as the settlers moved west.

ABOVE: *Bozeman, MT*

BELOW: *Greenville, IN*   OPPOSITE: *Albany, VT*

LEFT: *Bixby, MO*

BELOW: *Washington*

A later development in roof design was the gambrel, which was sometimes called a hip roof. The gambrel roof was developed as a way of creating more usable room over a given floor area, and the practice spread quickly among barn builders everywhere.

BELOW: *Northern Illinois*

ABOVE: *Flinthills, KS*

RIGHT: *Idaho*

Lancaster, PA

When our forebears were ready for a new barn, they didn't build it—they "raised" it. The favored structural material was timber, which was essentially squared-off tree trunks, a source of which they had an almost endless supply. The favored pattern was the bones of an animal, an example with which almost every farmer was intimately familiar.

Raising a barn was much more than a single man and his sons could manage, so helpers would come from far and wide. The Amish in Pennsylvania still follow the old traditions, both in the architecture of the structure and the communal effort to raise it.

Each structural part of the barn has a name. The sills are at the bottom, on the foundation. The verticals are posts. The space between each pair of posts is called a bay, and specialized elements are added here. For instance, in a cattle barn a space might be left for dropping hay down, so this would be a hay bay. An extension such as a shed might be added, and this would be an extended bay. If a window is pushed out, it would be called a bay window.

The timber over the opening at the near end is the door header. The top timbers are plates, the intermediate horizontal pieces are girts, and the diagonals are braces. A full-timbered barn was enormously heavy, which was necessary if it was to stand against the elements.

After the pieces had been cut and shaped, the cross sections of the whole structure would be put together by first laying them flat on the ground.

Each section was called a bent. After the first bent was raised into position, the bottom ends of the corner posts would be shaped into tenons. Each was then set over the mortise that was cut into the sill to receive it. Then the men would push the bent upright using long poles, often with ropes to help guide and steady the bent. Each successive bent would be raised and locked into place by the plates on top. A master barnwright was so practiced in his craft that the pieces of his barns were often interchangeable.

A fully timbered barn with mortise and tenon joints secured by pegs was made entirely of wood. There was a certain amount of flexibility in this form of joinery so that the barn could move with the stresses of wind and season much as a tree would move. When wrought-iron strapping was used to strengthen a barn at the joints, the rigidity at those points would break the timbers and, eventually, the whole. An all-wood barn like the Amish build may last three hundred years.

Lancaster, PA

Timber framing began to give way to milled lumber after the Civil War. Several factors contributed to this epochal change. Timbers were cut from the great trees of the virgin forest, but they were too large to move to building sites in cities. At the same time, the rise of industrial technology led to the sawmills of the lumber camps, which moved with the operations of the timber barons. Dimension-cut material began to dominate the market and provided a standardized vocabulary for builders—the two-by-four and two-by-six sticks and the one-by-eight boards.

The Chicago fire of 1871 destroyed 17,450 buildings, and the old hand-made ways were not sufficient for the immense task of rebuilding the city. This is where framing with milled, dimension-cut lumber first caught the public's attention. Most people had grown up with post-and-beam barns, and when they saw the spidery two-by-four framing of the Chicago building boom, they thought that these efforts would blow away in the first wind—they called it "balloon framing." It proved stronger than that and soon spread to the countryside. The age of post-and-beam framing was over.

Orange, VA

Nails have only been round in shape for a small part of their existence. The early European settlers held their buildings together with wooden pins, but as technology advanced into the New World, metal fastenings soon proved their usefulness. It began as a do-it-yourself process; traveling tinkers and ironmongers sold straps and rods of iron, and the local blacksmith or the farmer himself would cut nails from this stock as needed. This produced a flat-sided "cut" nail.

This helped solve another long-standing problem of barn builders: doors tended to work loose as they were used; even if they were made double-thick, gravity pulled at them, and all that banging to and fro loosened their boards. So nails would be made that were an inch (2.5cm) or so longer than the door was thick, and when the nail was driven through, the end would pro

trude out onto the other side. The builder would hammer the nail down flat (a process called clenching), which tightened the boards together. When a nail was clenched over the boards, it was said to be dead, hence the expression "dead as a doornail."

ABOVE: *Union City, IL*

BELOW: *Bozeman, MT*

The saltbox design was a favorite among settlers. The builder would put the long roof and lowest wall on the north side of the building; that would be the cold side, away from the sun, so it would have no drafty large windows, and the snow would stay unmelted on the roof and provide insulation. The opposite side would be facing south, so it had a high wall to let sun in the windows for light and heat.

This type of barn was also a lesson in practical orientation: the long roof of the saltbox pointed north, and most storm winds came from the west, or toward this viewpoint. Thus, the farmer would place an extra course of shakes on the far side of the roof, the west side.

ABOVE: *Washington*   OPPOSITE: *West Milford, WV*

BELOW: *Genesee, ID*

~

If there is one signature that Americans associate with a barn, it is the cupola. The structure and the word began in Italy, where the stem of the word meant "tub" and entered English usage as "cup." The Italian device was a decorative piece on top of a building that might enclose a bell. With wider openings and called a belvedere, it provided a view of the scene below.

Most Americans think that cupolas on barns are for ventilation. Oddly enough, when cupolas first appeared in rural America, they served no function at all. A cupola was the place where a builder could loosen up a bit after the stern practicality of the stalls and pens down below: it was the place where he could show off. So the cupola was a bit of vanity, useful only to the enterprising bird.

OPPOSITE: *Bournville, OH*

As barns and farms grew larger, barns began to burn down more often, and not always from lightning or an occasional overturned kerosene lamp. These fires were due to large piles of damp hay that did not dry out well; heat was generated deep inside the pile and, given the right conditions, flames broke out in spontaneous combustion.

The solution was more careful drying in the field and better ventilation in the barn. The cupola was already in place on many barns, a tidy little structure with its own roof and slats to keep out the elements. So farmers knocked a hole under their cupolas to turn them into ventilators. The decorative qualities of cupolas endured, however, and they continued to grace the skyline until sheet-metal workers came up with something that probably gained in efficiency but certainly lost in character.

ABOVE: *New Lebanon, NY*

BELOW: *Sonoma, CA*

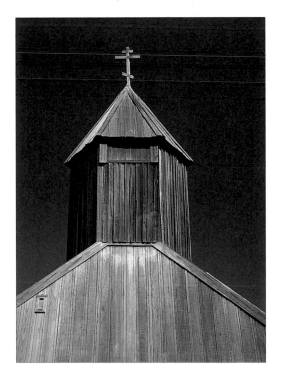

ABOVE: *Dutchess County, NY*

BELOW: *Lincoln, NE*

RIGHT: *Colorado*

LEFT: *Washington County, IA*

ABOVE: *Addison County, VT*

ABOVE: *Middlebury, VT*

B arns used to be dark inside, and there is little evidence that builders tried to make it possible to let in more light. It must be said, as always, that we look at such things through spectacles ground to the prescription of our own experience. We are accustomed to floods of artificial lighting everywhere, but an eighteenth-century yeoman had never seen a man-made light brighter than a flame. He hadn't seen a brightly lit interior in any other building, and it wouldn't occur to him that he should have such a convenience in his barn.

Then someone left an opening over the door so that when the door was closed, it didn't get entirely dark on the inside. This evolved into a long open space over the large door, which would be covered in the winter to keep out the bad weather. Soon, someone hinged the board so that it wouldn't have to be kept track of in the warm season. When glass became cheap enough, the board was replaced with glazed windows, a detail that soon spread from the barnyard to the houses of colonial gentry in the form of the fan lights over their doors.

RIGHT: *Woodstock, VT*

The earliest barn hinges were made of wood, in a simple two-part design: there was a stout pin fitted into the jamb, and a strap on the door with a hole on the end that dropped over the pin. The design crossed the ocean to the New World, and, making a name of the design, Americans called it a pintail hinge. As the design spread, the name was rounded off to pintle, and blacksmiths copied the design in wrought iron.

RIGHT AND ABOVE: *Putnam County, NY*   OPPOSITE: *Sweden*

Any barn that is meant to shelter hay will probably have a door at the top of the gable end and a protruding peak on the roof. Inside, there will be a steel trolley rail running the length of the roof peak with a set of tongs suspended from it by small wheels. The tongs are lowered to lift hay off a wagon on the ground. A system of ropes and pulleys then lifts the load to the peak and rolls it along the trolley to be dropped at the right place to build a mow of uniform depth. This addition began in the West, and it was big-job technology, as befitted the harvests of the Plains states.

RIGHT:
Brown County, IN

BELOW: *Overland Park, KS*

LEFT: *Kansas*

ABOVE: *Central Tennessee*

The earliest American barns didn't have doors; there was a large opening in the side, and when bad weather came, it was covered up with boards.

The first requirement for a barn door was size: it had to be large enough to let loaded wagons pass through. This led to problems: if a large door was on hinges, a correspondingly large amount of room was required to swing it open. More to the point, the door would be so heavy that the hinges of the day might pull loose from their moorings. So a method had to be found to support it from the top and slide it sideways.

The first solution was to fashion a rail over the door that resembled a stout gutter running twice the width of the opening, with a matching but inverted rail on the inside top of the door. The door piece fitted over the wall piece, and, with the application of enough grease and enough effort, the door could be pushed sideways.

The perfect solution had to wait until about 1840, when the railroads solved a similar problem in the design of their freight cars and came up with a steel strip above the opening and wheels on the door that fit over the strip. With the wheels set in lubricated bearings, the system would work indefinitely. It hasn't needed improvement to this day.

New York

OPPOSITE:
Owosso, MI

BELOW: *Kentucky*

ABOVE: *North Dutchess County, NY*

OPPOSITE: *New York*

ABOVE: *Northern Illinois*

RIGHT: *LaSalle County, IL*

Some barn doors were made to fulfill simpler requirements. One such need was to restrain the smallest residents on a farm, either to keep them in or keep them out. A widespread example is the "Dutch" door, with a lower half to close in the interests of restraint, and an upper half to open for light, ventilation, and society.

LEFT: *Oregon*

Columbiana County, OH

Round-topped doors were hardly ever found on barns. The builders avoided unusual forms because they were difficult to make and served no purpose that would not be equally well served by a square door.

ABOVE: *East Albany, VT*

LEFT: *Ruby Valley, MT*

The little roof over the door is called a penthouse. A penthouse could be small, just big enough to keep the dripping water from going down a farmer's neck when he stepped out the door.

RIGHT: *West Hurley, NY*

LEFT: *Oregon*

ABOVE: *Addison County, VT*

ABOVE: *Tracy, CA*

There are two things that we assume will be red: fire engines and barns. Whatever the origin of red fire engines may be, red barns began in the colonies.

Our forebears generally had a far keener understanding of the properties of their natural surroundings than we do. For instance, Antonio Stradivari, the legendary Italian violin maker born in 1644, treated the wood he used with a mixture of water, volcanic ash, and egg whites; no modern techniques have ever matched the sound of his instruments. Similarly, women in the New World made dyes of remarkable delicacy and permanence out of plants, and used them to color the threads for their needlework and textiles.

For finishing their barns, the settlers combined naturally occurring iron oxides, iron filings, lime, and milk to make their red paints. Red was the easiest color to make, and it was durable. The habit of painting barns red has survived even though the ingredients may have changed. Of course, not all barns are red. Now that durable paints can be found in an array of colors, barns may be blue, white, orange, or gray in addition to red.

LEFT: *Columbia County, NY*

LEFT: *Putnam County, NY*

BELOW: *Boxley, AK*

H ex signs were first painted on barns built by the Pennsylvania Dutch about a generation into the nineteenth century. Most Americans like to think that hex signs were used to either encourage the intervention of rural spirits or keep them away. An equally likely explanation is that these pleasing decorations were actually invested with no mystical content at all—the farmers just liked the look of them.

Round barns enjoyed widespread favor in America between about 1895 and 1915. This was partly due to the example set by the Shakers, who were admired as prudent, hard-working farmers and who also favored round barns for reasons inherited from Athens, Greece in the fifth century B.C. The philosopher Plato believed that the perfect geometric form was the circle and, like all other forms of perfection, it could be found only in the mind of God. Applying the circle to practical building appealed to the religious Shakers.

The circle may be a perfect and inspiring form, but it is very difficult to make a flat door fit tightly on a curved wall, and even harder to make a curved window.

Wisconsin

OPPOSITE: *Polk County, WI*

One solution to the problems of flat openings in a curved wall is to not have any openings. Another solution is to make a barn sort of round, as in the twelve-sided models often seen in the Midwest.

BELOW: *LaSalle County, IL*

The rounded roofline evolved from the gambrel roof and had the same advantage: it provided more volume over a given floor area than the straight planes of a gable roof.

The bow truss roof is a rather recent development that became popular in the 1930s and eliminated the need for most of a barn's interior bracing beams. A bow truss roof was quite difficult for a farmer or even an experienced builder to make. However, like almost everything else a farming family needed, a complete precut bow truss barn could be ordered from the Sears & Roebuck catalog, in various sizes and modular variations to suit different farming needs.

ABOVE: *Custer County, NE*

BELOW: *Idaho*

BELOW: *Eastern Washington*

Farmers have always worked hard, but they don't like to suffer needlessly any more than any other mortal. So, as they learned the uncertainties of New England weather, they soon looked for ways to avoid those long walks out to the barn during a blizzard.

The solution proved to be a happy meeting of necessity and convenience. When farmers needed more rooms on the farmhouse, they would build them between the barn and the house until they eventually bridged the gap. This led to the "big house, little house, back house, barn" tradition of add-on architecture that became characteristic of the northern New England farm.

ABOVE: *Ludlow, VT*

LEFT: *North Branford, CT*

For most of human history, mass communication meant raising your voice. Alternatives were eventually found, one of the most engaging was barn art, and the promises therein would pop the eyes of farmers and children alike.

The most modest of these might remind the men of their favorite chew. Still in the same mercantile territory, a handsome painting hoped to remind passers-by of the desirable qualities of cigarette rolling paper, and it's unusual because there is no lettering. By this date, a picture was evidently worth at least a dozen words.

Other barn paintings were more ambitious in theme and scope, and one of the most awaited paint jobs announced the annual arrival of the circus. An advance man would make his way through the countryside and arrange to rent the wall of a prominent barn. The painters followed, and soon tigers, clowns, elephants, and trapeze artists sprang to vivid life beside the cornstalks and potato fields. The name of the circus was prominently displayed along with the dates that it would be nearby. If a larger population and suitably located roads indicated the need for greater exposure, the advance man might rent two walls of a barn, or even all four of them.

The big touring "canvas" shows are now only a romantic memory, but the painted barns that proclaimed their glories linger on in our language if not along our country roads: when a politician's campaign manager plans a maximum effort, he tells the staff to get ready for a month of four-walling.

Gursey County, OH

ABOVE: *Garfield, WA*

BELOW: *Southern Michigan*

Tobacco was a major cash crop and called for a highly specialized barn for the essential process of drying and curing. Air and space were the only necessities, and all tobacco barns had some way of letting air through the walls. In the South separated logs or holes were often utilized. Up north, movable wall boards were used. Alternate planks were hinged at the top and, during the drying season, they were propped open, an adaptation that made the barns look as though they were falling apart.

Southern curing barns sometimes used a slow-burning fire in the middle. Wet sawdust was sprinkled on the hardwood fire to keep it slow. The earliest small-farm barns would have tobacco leaves nailed to the roof boards to dry, hence the startling number of nails found in some old barns. All later tobacco barns have tiers of poles, and tobacco sticks are dropped over them.

BELOW: *Westin, MO* OPPOSITE: *Westin, MO*

When the first European settlers arrived in America, they found the natives using a food staple that had never caught on in the old country: corn. This splendidly adaptable food must be dried carefully or it will rot. It must be exposed to as much air as possible but sheltered from the rain. Given the lessons of solid geometry, farmers would build several small corncribs rather than one large one, and they would often roof them over in pairs so that the wagons could be brought in between them to unload.

A major enemy of corn is rats, and the cribs were built up off the ground so that rodents couldn't get into them as easily. A determined rat could still climb the short poles on which the cribs sat, so the farmers would either cap each pole with an inverted pie tin that the rat couldn't get around or they would wrap the pole with tin

that was too slippery to climb. Eastern farmers almost always built their corncribs with walls slanting inward for more protection from rain. Farmers in areas with less rain could indulge the easier carpentry of making a square wall.

OPPOSITE: *Southern Iowa*

Buffalo County, WI

What the church steeple is to a New England town, silos are to the farms of America. However, silos are a rather recent page in our national picture book. The first round, outside silo was probably built by Fred Hatch, in McHenry County, Illinois, in 1873.

Farmers have always needed to store feed for their cattle, and if it was placed in a silo it was called silage. Anything that could be eaten could be used for silage: grass, oats, or corn. Since each acre (0.4ha) of an average farm would produce a greater volume of corn than any other crop, corn was preferred for silage. The entire plant could be chopped—stalks, leaves, cobs, and kernels—and all of it dumped into the silo.

Unlike hay, silage had to be kept away from air, so a large, tight container was needed. The first solution was a tall square-cornered bin built into a corner inside the barn. These

bins were difficult to keep air-tight, and as air leaked in, the contents went sour. So the silo was moved to the center of the barn and constructed on a circular plan.

Midway through the eighteenth century, the emerging

American iron industry created a huge demand for wood to be turned into charcoal for iron making. Farm builders turned to stone silos, and to make them airtight, half the total height was usually below ground level. Brick silos were the next natural

Virginia

step, and, since the combina-
tion of brick and masonry was
almost airtight, the whole
structure could be above
ground and outside the barn.
When wood became plentiful
again, the outside wooden silo
found its familiar form.

However fitful silo development may have been, the building technique that was finally
adopted was a very old one. A
silo is essentially a very tall
barrel; the wooden siding
boards are of different lengths,
so the joints come at random
intervals, and as each level is
laid up, an iron band is tightened around it.

Eastern farmers were not
quick to adopt the idea of tall
round silos, and it took this
perfected design twenty years
to take hold in the seaboard
states. Ironically, tall wooden
silos are better suited to the
East than to the Midwest. The
problem is in the boards, or
what are called staves in a bar-

Allen County, IN

rel. The wind of the prairies
puts a steady pressure against
anything that gets in its way,
particularly tall things. So an
aging silo, growing loose in its
joints, will gradually lean away
from the wind like a tall ship
heeling to leeward.

Silos built in our own day
are often made of metal and
have elaborate systems of pipes
and blowers to load the raw
silage into the top.

LEFT: *Rainier Valley, WA*

RIGHT: *Lewiston, ID*

BELOW: *Minnesota*

LEFT: *Craftsbury Commons, VT*

OPPOSITE: *Ripton, VT*

Here, one of mankind's oldest, simplest, and most useful inventions has become art. That graceful interlock of curves hanging in the gable of this trim barn is the yoke for a single ox. "Yoke" also means a pair of oxen working together.

Early farmers preferred oxen to horses for at least one compelling reason. the bull for breeding, the ox for pulling, cattle for meat, and cows for milk all belonged to the same family of animal.

The ox was also notable for its narrative usefulness. One of the oldest examples of fatherly advice came from a king of Heracleopolis to his son, in 2100 B.C.: "More acceptable is the character of the straightforward man than the ox of the wrongdoer." In biblical times, the ox was considered so valuable that it took its place in the Ten Commandments: "Thou shall not covet thy neighbor's house, nor his wife, nor his manservant, nor his maidservant, nor his ox." Less sternly but still in the pre-Christian era, the poet Horace found a place for this noble beast in his own lyrical vision of virtue: "Happy is the man who far from schemes of business, like the early generations of mankind, works his ancestral acres with oxen of his own breeding, from all usury free."

Brownsville, VT

"Man may work from sun to sun, But woman's work is never done."

The origins of that homily are lost in time, but whoever first said it never lived on a dairy farm, where most work is done by men. Milking cows can be neither hastened nor delayed, and it has to be done every evening, often after the sun has set.

The list of chores also includes transferring the milk to cans for shipment to market, paying strict attention to washing all the containers and implements, pitching hay, setting out feed for the animals, cleaning the barn stalls, hauling manure, and tending to sick animals—a rural litany that would convince any dairy farmer that his work, too, is never done.

BELOW: *Marlboro, MA*

OPPOSITE: *LaSalle County, IL*

A cow's diet varies widely: during warm weather the animals will be set out to pasture and graze on fresh grass, while at other times they will be in a feed lot or in the barn. Farmers also put their cows into a cornfield after the crop is in so that the cows can clean up the farm before winter by eating the husks and stalks left by the corn pickers.

These changes in diet are immediately evident in the taste of the milk, and people who grew up in households where the milk was supplied straight from a nearby dairy farm could always tell when the flock of cows moved from the barn to the pasture. In fact, an educated palate could even tell when the animals had gotten into a field not meant for them; for example, if they'd been in the onion patch for a few hours, the milk they produced would immediately betray their escape from their usual grassy pasture. Today, all milk on the market is a blend made from many farms and these local idiosyncrasies are lost.

BELOW: *Colfax, WA*   RIGHT: *Peacham, VT*

ABOVE: *Putnam County, NY*

The next stage in hay making was the baler, and the bales would often be lifted into the barn with some variant on an endless inclined belt.

RIGHT: *Putnam, CT*

Highland County, VA

In older times, hay for the animals would be pitched onto wagons, taken to the barn, and pitched into the hayloft. This had a social as well as nutritional dimension: when a hay wagon passed, children used to consider it a sign of coming luck if they could pull off a handful of its load, though their many haymow games probably didn't please the farmer.

ABOVE: *Missouri*

Newer harvesting machines pack the hay into huge rolls for transport that are often stored outside. The latest device is an enormous self-powered micro-wave oven, which is towed through a hayfield; cut grass goes in one end and dried hay comes out the other. Farmers no longer need to make hay while the sun shines, and children must now find new omens and games.

RIGHT: *Missoula, MT*

≈

PART II

FUNCTION

Cato the Elder lived two centuries before Christ, and the lessons of working the soil were already well understood: "A farm is like a man," he wrote. "However great the income, if there is extravagance but little is left."

Then and now, farmers have understood that their buildings and tools should be sufficient to the task—nothing more and nothing less. This joining of form and function is at the heart of the many pleasures of the barn.

OPPOSITE: *Waitsburg, WA*

≈

This trim fortress was made by a stonemason at the height of his powers. He framed the corners with large cut blocks and made the walls of mortar and field-stone, a job complicated by the many openings for doors and windows. The whole was fin-ished with smooth masonry fac-ing, though much of it has now weathered away.

The most interesting details blend with the shadow of the tree on the near gable end. A barn as tightly made as this one would have needed extra venti-lation, so the mason made ten vertical slots in the wall. This technique flourished in late-eighteenth-century America, and it dated back to conven-tions used in much earlier cas-tles. These slots were called "loops," and the most ingenious part of the design was on the inner wall, where the narrow outer opening widened at a sharp angle.

This was crafty engineering: the expanding cross section would create a venturi effect whenever the air was moving, an early form of forced ventila-tion. In medieval Europe, the slots were called "wind eyes," which gradually became "win-dows." Not much light got in through a wind-eye, so a farmer would whitewash the inner sur-face of the thick wall at the loop, thus amplifying the light with a reflecting plane.

LEFT: *Southwestern New Jersey*

Pennsylvania

What classical half-timbered architecture was to Tudor England, this eye-catching beauty must surely have been to its makers' community in Pennsylvania. They began with the posts and tie beams, purlin braces and corner braces, and the door header and hood of a post-and-beam barn, but instead of sheathing the barn with planks, they filled in the remaining spaces with fieldstone. The top hood shelters a door for the hay trolley at the peak of the gable and another below it for the middle floor. Instead of a sliding door, which was in widespread use by this time, they used swinging doors on pintle hinges.

Like a noble ruin preserved from the days of ancient Rome, this stone wall is all that remains of one of the largest barns ever built. It's the North family barn, the work of the Shaker farming community in Lebanon, New York, where it was built soon after the Civil War. The Shakers combined the most traditional religious austerity with the most forward-looking elements in their farming practices. This barn was built entirely of stone, a building material that had been plentifully supplied by the glaciers. Average workmen could do everything needed to raise a wooden barn, but highly specialized skills and unusual lifting techniques were required to build with stone. Thus, while stone barns were almost impervious to decay and were a long-lasting investment, they never became as popular as wooden ones. This barn was nearly three hundred feet (91.5m) long and five stories high. It burned in 1972.

Lebanon, NY

ABOVE: *Near Kansas City, MO*

A thrifty builder would use whatever material was at hand. For example, there are vast beds of limestone in Illinois and Indiana. This barn sits securely on a stone foundation which also forms the walls of the ground-level spaces.

OPPOSITE: *Konza Prairie, KS*

This superb piece of work almost transcends our understanding of what a barn can be. This barn, built in the mid-1820s by the Shaker community of Hancock, Massachusetts, was 270 feet (82m) in circumference and twenty-one feet (6.5m) high at the eaves.

All farms are to a greater or lesser extent a system, that is, they are a collection of separate parts dedicated to a common end. The Hancock barn is a system all in itself: it was designed to accommodate fifty-two cows in stanchions on the main floor facing inward toward the haymow, which took the form of an open-sided central tower. The hay wagons were driven into the barn and up a wide ramp to load the tower from the top, and the hay was pulled out as needed down below. This arrangement greatly reduced the effort needed in the endless job of handling hay and

Hancock, MA

required only the initial, and rather minimal, effort of pitching the hay down from the wagons onto the upper section of the tower. Gravity did the rest.

The stone walls of the Hancock barn were 3½ feet (1m) thick and almost perfectly flush on both the inside and outside surfaces. The planking in the floor had to be tapered to perfect the radial fit. The rafters seem light as an Oriental parasol, but they sustain the great weight of the snow in a New England winter.

RIGHT: *Puget Sound, WA*

BELOW: *Hancock, MA*

RIGHT: *Garrison, NY*

Hancock, MA

No amount of artifice will eliminate one of man's oldest occupations. It's no accident that one of the twelve labors of Hercules was cleaning out the Augean stables, where the king of Elis kept a herd of three thousand oxen.

The Shakers characteristically amended their plans as they went along. The wide floor was initially intended as a threshing floor for the preparation of grain, but this use was set aside when they realized that the purposes of threshing did not efficiently coexist with those of dairying. The place burned, and, undaunted, the Shakers stole a march on Hercules and rebuilt the barn with improvements: a new lowest level was established so that the manure could drop down from the stanchions and be collected in wagons that were driven through underneath.

Nor was that the end of the good brought by the fire's ill-wind. The Shakers realized that the enormous barn needed more light and air, so they replaced the original conical roof with the twelve-sided arrangement that exists today, and added a cupola of matching symmetry. This avoided the difficulties that round framing always brought, and is another example of the Shakers' equation of good in simplicity. Even the picket fence reflects a sense of coherence and beauty; instead of the conventional straight line at the top, the pickets are cut in a curve that reflects the larger forms of the barn that they protect.

BELOW: *Hancock, MA*

A country builder was at the top of his form here, and he left this wonderful monument. One of its classical links is deception; like many masterworks of European design, it has dimensions that are difficult to judge. Actually, its size far surpasses the ambitions of most barn builders: it is sixty-five feet (20m) tall, the size of a six-story commercial building.

This farm graced Vermillion County, in western Indiana, but its barn served another emerging need of the day. In an age before electronics, fliers navigated by whatever means they could: rivers, railroads, roads, or large buildings. Few buildings could have been larger or easier to spot than this one, and it became a marker for commercial pilots on the air route between Chicago and Terre Haute, Indiana.

Today, the cupola is crumbling, and the wind is tearing

Vermillion County, IN

the shingles away to leave the underlying rafters exposed to the snow and the rain. Unless some patron of the arts finds this barn, gravity and the slow fires of time will destroy it.

The extraordinary compounding domes of this famous barn have graced the Manchester farm for five generations.

The date so prominently displayed on the name plate provides an important insight into the thinking behind it. A powerful faith in progress swept across America at the end of the nineteenth century, and the conviction spread that the com-ing years would bring some-thing close to utopia; round barns were seen as both the expression and fulfillment of this new dispensation of virtue.

Thus, the Manchester barn was built with the silo in the middle, under the substantial cupola. Hay and feed were in the center, with the stock arranged on the circumference as in the Shaker design of eighty years earlier. In fact, one enthusiast went so far as to the-orize that this was the realiza-tion of a higher plan, since cows were wedge-shaped, they could be fitted into a round barn head inward for more effi-cient feeding. In this, however, he might have been pushing the idea of perfection a bit further than the realities of bovine anatomy would sustain.

New Hampshire, OH

If this handsome barn was built today, we would say that the farmer bought the catalog. The barn was constructed on a fitted stone foundation, and the center section is made of brick. The center door has a rare rounded top, made possible because it is very difficult to make arches out of wood but quite easy to make them out of brick. The facing wall also displays the way masons provided ventilation: they included decorative panels of open brickwork for air-flow spaces, the mason's version of the wooden louver at the top of the right-hand section. The small door at the left was made in two sections so that the upper part could be opened for light and air while the lower section was closed to keep wandering four-footed creatures either in or out, as the situation required.

As if that weren't enough, this barn also has something that is almost never seen: decorative trim that serves no practical purpose at all. Not only that, but the shadow shows that it is open-work decoration.

Weston, MO

ABOVE: *Nova Scotia, Canada*

BELOW: *Shoreham, VT*

Equipment suppliers liked their customers to display tin advertising panels like these. The ones shown here (bottom) testify to the farmers' trust in the Girton milk storage tank and the Surge milking machine. Milking machines changed all aspects of dairy farming; a farmer working by hand could only milk seven cows an hour, which is why dairy farms had to wait for milking machines before they could go to large-scale commercial production. But this led to the larger problem of refrigeration. Large amounts of milk could not be handled by the traditional milk shed that was cooled by water from the farmer's spring.

The most important requirement of barns is that their form follow their function—or else. This builder tried a variant (right) and paid a price. By joining two gable roofs side by side, he created a valley and thus had to drain away the double dose of precipitation that would collect where they met in the middle. The eaves on the right side were evidently placed above a place that would not be helped by rainwater running onto it. So the builder was forced to put a gutter there, and then ran it diagonally down to a junction with the downspout from the valley in the roof and on across to the far side. It is a rather complicated arrangement by contrast with the two simple siding forms: the boards on the right wall are put on flush, while the boards on the gutter wall are tightened by nailing battens over the cracks between the boards.

ABOVE: *Muscatine County, IA*

RIGHT: *Northfield, MA*

The first Mount Hermon on record is documented in the Bible, rising above the cedars of Lebanon and the Sea of Galilee. The shingle pattern in this barn (above) proclaims that this is Mount Hermon School, in Northfield, Massachusetts. It was established in 1881 with an emphasis on spiritual values as well as the requirement that all students work on the school farm to understand one of the other nourishing dimensions of their life. The message on this barn, however, had to wait until the appearance of colored asphalt shingles.

ABOVE: *Grand Teton National Park, WY*

These structural techniques show variants that were more common in the West than in the East. Since tall, thin softwood was plentiful in the western states, westerners made barns that were essentially log cabins chinked with mud or plaster for tightness. These two barns in front of the Grand Teton Mountains of Wyoming exemplify another tradition; they are both quite similar in design and were probably the work of a builder who brought his services to the whole valley.

ABOVE AND RIGHT: *Grand Teton National Park, WY*

No map or atlas is needed to locate this barn. The vast area of the flat shed roof would never be able to stand up to the winter weights of snow country; this barn is on the warm Sonoma coast of California.

LEFT: *Sonoma Coast, CA*

This farmer started with a basic gable-roof barn. Needing more storage space, he built a skirt of sheds all around it. Now, forsaken in the Missouri National Grasslands, it is being reclaimed by the weather. Before long, it will fall to the ground.

RIGHT: *Little Missouri National Grasslands, ND*

This handsome barn (below) in Torrington, Connecticut, was made so well and cared for so faithfully that it will probably outlast many future generations of owners.

But if the boards of a barn loosen and the shingles blow away, the bad weather gets in and the wood begins to rot. Then it is only a matter of time before gravity takes over; even the best of barns will return to mulch and earth. Given enough time, the old bones may fertilize a new forest and perhaps make new barns in a distant return to the old ways.

BELOW: *Torrington, CT*　　　RIGHT: *Grand Teton National Park, WY*

ABOVE: *Eastern Washington*

LEFT: *Washington*

BIBLIOGRAPHY

Klamkin, Charles. *Barns: Their History, Preservation, and Restoration.* New York: Hawthorn Books, 1973.

Sloane, Eric. *An Age of Barns.* New York: Funk & Wagnalls, 1967.

_____. *A Reverence for Wood.* New York: Funk & Wagnalls, 1965.

Ziegler, Philip C. *Storehouses of Time.* Camden, Maine: Down East Books, 1985.

In addition to the many books available on barns, any state historical society should be able to provide plenty of information on these architectural treasures.

PHOTOGRAPHY CREDITS

© Amstock/Kenneth Martin: 31 top, 32 bottom, 78, 82 bottom, 91, 94, 95 left, 96, 97 bottom, 114

© Christopher Bain: 32 top, 35, 37, 40 top, 40 bottom, 55 right, 56 left, 56 bottom, 57 top, 76, 82 top, 101 top, 101 bottom

© David Brownell: 63 top

Dembinsky Photo Associates

© Ian J. Adams: 30, 98, 99; © Gary Buss: 48 left; © Willard Clay: 21 bottom, 50, 112–113; © Daniel Dempster: 18 bottom; © Terry Donnelly: 51, 61; © Gerard French: 62 left; © Michael Goldman: 49; © Darrell Gulin: 28 bottom, 62 right, 73 bottom; © Michael Hubrich: 45; © G. Alan Nelson: 34, 47, 60, 102–103; © Carl R. Sams: 65 bottom

FPG International

© James Blank: 104; © Paul Boisvert: 41; © Dick Dietrich: 86; ©Kenneth Garrett: 70; © Jeffrey Myers: 117; © David Noble: 115; © Richard T. Nowitz: 5, 55 left; © Jerry Sieve: 53; © Stephen Simpson: 18 top, 27 bottom; © Ulf Sjostedt: 41; © Richard Smith: 27 top, 75; © Ron Thomas: 6–7, 8, 19, 23, 29, 33, 39, 63 bottom, 74, 77, 81, 105, 108–109, 110–111; © Jack Zehrt: 13; © Nikolay Zurek: 56 right

Leo de Wys

© David Burnett: 85; © Howard Dratch: 48 right; © Everett Johnson: 24, 83; © Brian King: 90; © Henry Kaiser: 54 right; © Stephen Simpson: 54 left

© Daniel Lyons: 25

Midwest Stock

© Jim Hayes: 42; © Michael P. Manheim: 103; © Kevin Sink: 2, 20, 22, 44, 66, 67, 69, 80, 92, 93, 100, 106, 116; © Ben Weddle: 97 top

© Susan Nash 1996: 3

Tom Stack & Associates

© Tom Algire: 16; © Terry Donnelly: 36, 43, 58 bottom, 68, 79; © Sharon Gerig: 71; © Bob Pool: 72; © John Shaw: 21 top, 28 top

Tony Stone Images

© Doris De Witt: 59; © H. Richard Johnson: 10; © Claudia Kunin: 84 top; © Lois Moulton: 65 top; © Greg Ryan & Sally Beyer: 57 bottom; © Paul Sisul: 73 top

Superstock

© Ping Amranand: 26; © Scott Barrow: 95 bottom; © Brent Cazvedo: 52 top; © E.F. Productions: 62 top; © David Harvey: 52 bottom, 95 top; ©Hank Miller: 31 bottom; © John Warden: 64

© Allan Weitz: 58 top, 88–89